Name _____   Date _____

`MW01517167`

INTRODUCTION TO MAGNETISM

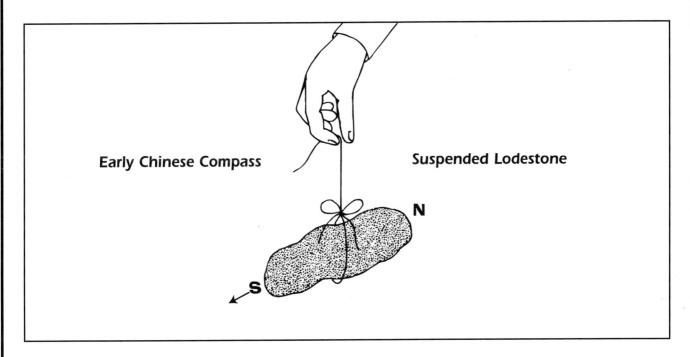

Early Chinese Compass   **Suspended Lodestone**

N

S

The ancient Greeks and Chinese knew about magnetism over 2000 years ago. They learned about magnetism from exposed pieces of black iron ore. This ore, which we call magnetite today, gave people direct contact with magnetism. Almost 2000 years ago, the Chinese hung thin pieces of this ore so that it could swing freely from a thread. The hanging stone made a simple direction finder. These natural, magnetic rocks, became known as lodestones or "leading stones." By the 1200s, sailors were using iron needles, magnetized by pieces of lodestone as direction finders instead of using the lodestone itself.

Today we know how to make strong, artificial permanent magnets. They are used frequently in our everyday lives.

List eight uses of magnets in or around your home or school.

1. _____
2. _____
3. _____
4. _____
5. _____
6. _____
7. _____
8. _____

MP3427 Magnetism and Electricity

Name _____ Date _____

WHAT DO ALL OF THESE ITEMS HAVE IN COMMON?

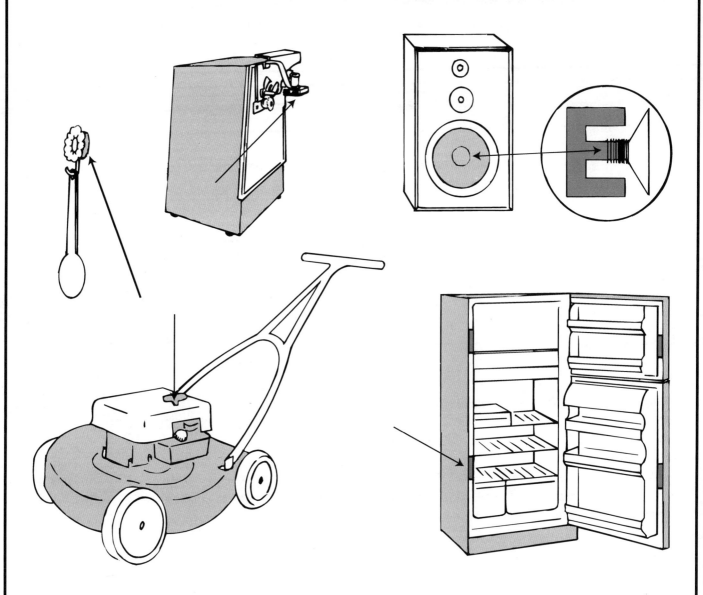

List these and other household items that use magnets. Explain how each uses its magnets.

_____

_____

_____

_____

_____

_____

_____

_____

2

Name _____   Date _____

MAGNETIC FORCE

Magnetism, like gravity, is a force that cannot be seen. Every magnet, however, has an area or space in which it exerts its force. This area or space is called the magnetic force field. The size of this field depends upon the strength and size of the magnet.

To help picture the presence of a magnet's field, try the following experiments.

ACTIVITIES

Place a sheet of white paper over a bar magnet. Sprinkle iron powder or iron filings lightly over the paper. Tap the paper gently. Draw a picture of what you see.

Repeat this activity with a horseshoe magnet. Draw a picture of what you see.

Each line or path pattern you see in these two experiments is part of a complete curve or loop. These are magnetic lines of force. Magnetic lines of force leave the north pole and enter the south pole of a magnet. A magnet is completely surrounded by these lines of force.

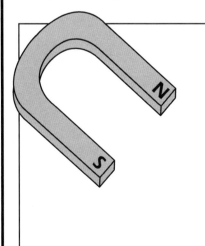

MP3427 Magnetism and Electricity

Name _____   Date _____

MAGNETIC POLES

All parts of a magnet do not show equal strength. The following activity will make it easy for you to find the poles on a bar or horseshoe magnet.

The areas of greatest strength or attraction on a magnet are called the poles. They are usually marked on a magnet with an **N** (north pole) and an **S** (south pole).

ACTIVITIES

Lightly sprinkle powdered iron over a sheet of paper. Then, move the entire bar magnet across the surface of the paper. Lift the magnet and observe where most of the powder clings to the magnet. Draw your results.

Repeat this activity with a horseshoe magnet. Draw your results.

MP3427 Magnetism and Electricity

Name _____ Date _____

BEHAVIOR OF MAGNETIC POLES

ACTIVITIES

Place two bar magnets about three inches apart as shown below. Slowly push magnet A toward magnet B.

1. Describe what happened to the magnets. _____
_____

Again, place the two bar magnets about three inches apart as shown below. Slowly push magnet A toward magnet B.

2. Describe what happened to the magnets. _____
_____

Now, repeat the above steps with two horseshoe magnets.

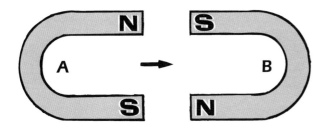 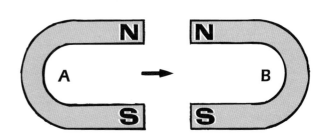

3. Describe what happened to each pair of magnets.

_____

_____

_____

You have observed one of the laws of magnetism in action. Unlike (N and S) poles attract each other. Like poles (N and N or S and S) repel each other.

Name _____ Date _____

MAKING ARTIFICIAL MAGNETS

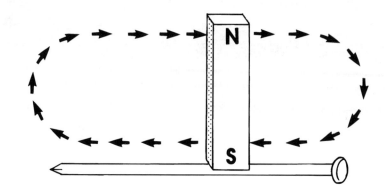

ACTIVITIES

Stroke a three-inch nail with a magnet. Each stroke on the nail should be in the same direction and with the same pole of the magnet. Stroke the nail 10 times in this way. Now, dip the nail into some iron powder. Observe how much iron is attracted to the end of the nail. Now, stroke the same nail 30 more times in the same way.

1. How much iron powder does the end of the nail attract now? _____
   _____

2. Describe how the number of strokes changes the strength of the magnetism in the nail.
   _____
   _____

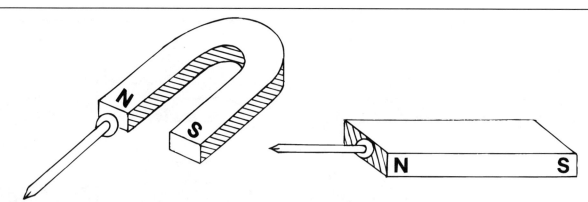

Place the flat head of a nail on one end of a magnet. While the nail is on the magnet, touch the nail to some iron powder.

3. Is the nail magnetized? _____

Allow the nail to remain on the magnet for about five minutes. Then, remove the nail from the magnet. Again, touch the nail to the iron powder.

4. Is the nail still magnetized? _____

The nail (soft iron) magnets you have made are examples of temporary magnets. Both nails will gradually lose their magnetism.

6

Name _____   Date _____

MATERIALS ATTRACTED BY MAGNETISM

Materials not attracted by magnets are called non-magnetic. Such materials cannot be magnetized. Only magnetic materials can be magnetized. The most common magnetic materials are iron and steel. Other, less common magnetic materials are cobalt and nickel.

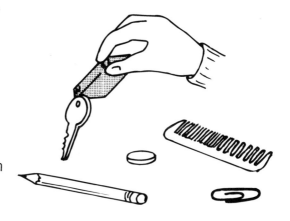

ACTIVITY

Gather a wide assortment of small items. Test each of these items to see if they are attracted by a magnet. Record your results on the chart.

Item Tested	Attracted by Magnetism	
	Yes	No
Paper Clip		
Chalk		

Item Tested	Attracted by Magnetism	
	Yes	No

7

Name _____   Date _____

EFFECTS OF NON–MAGNETIC MATERIALS ON A MAGNETIC FIELD

Non-magnetic materials allow a magnet's force field to pass through them freely without hindering or gathering in the lines of force.

ACTIVITIES

Wrap aluminum foil tightly around the end of a bar magnet. Dip this end into a pile of powdered iron. Observe how much iron is picked up. Remove the foil from the magnet. Next, wrap the end of a magnet with paper. Again observe how much iron powder is picked up. Repeat this activity with plastic wrap, cloth, and rubber sheeting.

1. List the materials that allowed the magnet force field to pass through them. _____

_____

Fill a saucer or plastic dish with about 1/2-inch of water. Stick a thumbtack into the center of a 1-inch square of thin styrofoam. Float the thumbtack head down on the water. Move a magnet around beneath the dish.

2. What happens? _____

3. Does the water appear to keep the magnet's force from passing through the dish?_____

_____

Place a thumbtack head down on a small piece of glass. Move a magnet around beneath the glass.

4. Describe what happens._____

_____

Repeat this activity using thin pieces of styrofoam, cardboard, or fiberboard.

5. Describe the results of each material used. _____

_____

_____

Repeat the last activity again. This time use a flat piece of metal cut from a can. Use the same magnet. Be careful of sharp edges.

6. Does the thumbtack move about as it did in the last activity? _____

_____

The flat piece of metal you used contains magnetic material which gathers in some of the lines of force from the magnet. This results in less pull on the thumbtack.

MP3427 Magnetism and Electricity

Name _____   Date _____

HOW A COMPASS WORKS

The Earth is a huge, powerful magnet surrounded by a powerful force field. A magnetized compass needle points to the North and South Poles of our planet. It is important to remember that the Earth's north and south magnetic poles are not the same as the Earth's geographic poles. The Earth's magnetic poles do not remain at a constant point as do the geographic poles. The Earth's magnetic poles have shifted position throughout its history and, at present, are well over 1000 miles from the Earth's geographic poles.

ACTIVITIES

Set up the following experiments in an area where they will not be disturbed. They should also be placed at a distance from each other and from other magnetic materials.

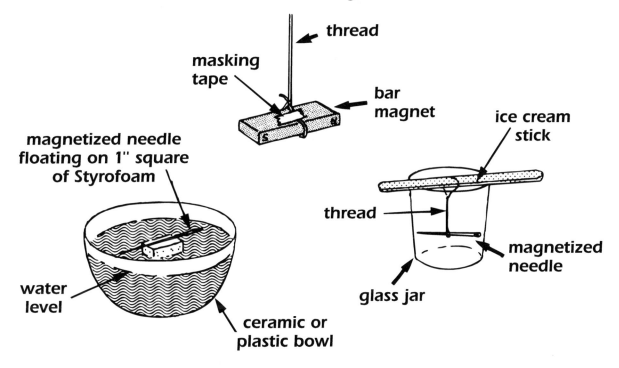

Allow each of the above activities to remain undisturbed for five or six minutes. Then, observe the direction in which the needles and magnet have pointed.

Record your observations. _____

_____

_____

_____

A free-swinging magnet or magnetized needle will always line up on a general north-south line. The needles and magnet in the above activities act as simple compasses.

9

Name _____ Date _____

THE EARTH AS A MAGNET

If a large bar magnet were placed beside the Earth, it would position itself with its south pole near to the Earth's north pole. The opposite poles would be attracted. The north pole on the bar magnet would be attracted to the magnetic south pole of the Earth.

What might happen if the bar magnet were reversed? _____

_____

_____

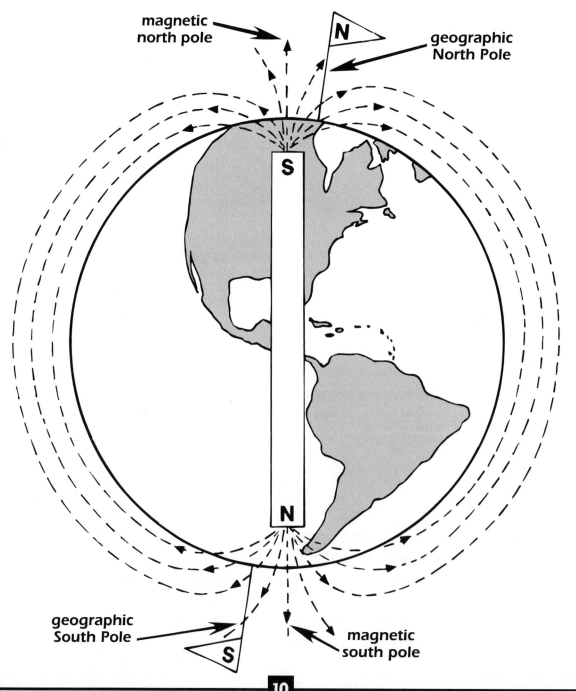

MP3427 Magnetism and Electricity

Name _____   Date _____

MOLECULAR THEORY OF MAGNETISM

All matter is made of tiny microscopic particles called atoms. Scientists believe that in magnetic materials groups of atoms called molecules act as tiny magnets. When the poles of these tiny molecule magnets are scattered, facing different directions, there is no magnetic field. When the north and south poles of the molecule magnets are lined up, facing the same direction, the material has a magnetic field. The magnetic strength increases as more of the molecule magnets are made to face in the same direction.

Heating a magnetized nail or beating it with a hammer will cause the nail to lose its magnetism. The heating and hammering both cause increased molecular movement within the nail. This causes the poles of the molecule magnets within the nail to again become scattered and lose their magnetic force.

ACTIVITY

Magnetize two 3-inch nails by stroking each nail 50 times with a magnet. Lay one of the nails aside. Grip the other nail in a pliers and hold the nail in a propane flame until the nail turns red. Then drop the nail into a container of water. Allow the nail to cool in the water for about one minute.

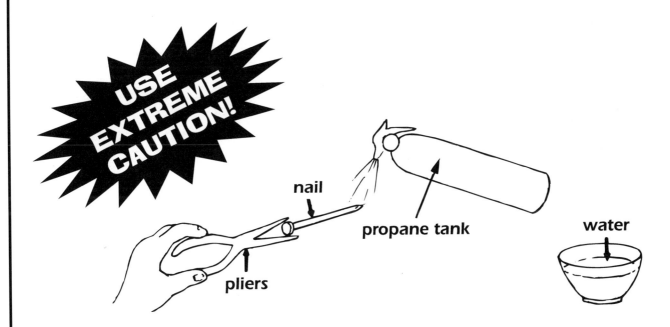

Test the magnetic strength of both nails by touching each nail to some iron powder.

1. How much iron powder does the heated nail collect? _____

2. How much iron powder does the non-heated nail collect? _____

3. Which nail shows the strongest magnetism? _____

MP3427 Magnetism and Electricity

Name _____   Date _____

ORDERLY MOLECULES

A MOLECULAR MAGNET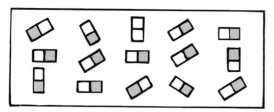

Unmagnetized molecules are not arranged in any particular order.

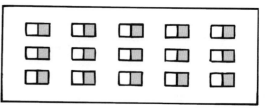

Magnetized molecules are arranged in an orderly manner.

SUPPORT FOR MOLECULAR THEORY

ACTIVITY

Stroke an iron nail with a magnet.

1. Is the nail magnetized? _____

_____

_____

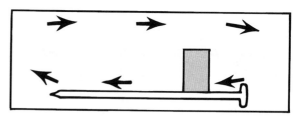

2. Stroke the nail in the opposite direction. What happens? _____

_____

_____

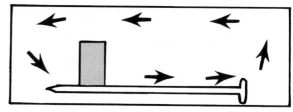

3. Are the poles reversed? _____

Draw a picture to show how the molecules would be arranged in the nail.

When an iron magnet is broken into pieces, all new pieces of the magnet will retain their polarity.

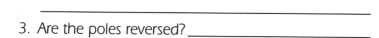

MP3427 Magnetism and Electricity

Name _____   Date _____

REVIEW OF MAGNETISM

Complete the following.

1. Magnetic lines of force are invisible. (True–False) _____

2. Magnetite is an artificial magnet. (True–False) _____

3. What are the magnetic poles of a magnet? Where are they located?_____

   _____

   _____

   _____

4. Magnetic lines of force form a complete loop. (True–False)_____

5. What is a magnetic force field? _____

   _____

6. Some (like–unlike) poles attract each other. (Like–Unlike) poles repel each other. _____

7. Non-magnetic materials can be magnetized. (True–False)_____

8. Describe two ways of magnetizing an iron nail using another magnet._____

   _____

   _____

   _____

9. Some non-magnetic materials will allow magnetic lines of force to pass through them freely.
   (True–False)_____

10. Heating a magnet will weaken it. (True–False) _____

11. A free-swinging, magnetized needle hanging on a thread is a type of compass.
    (True–False)_____

12. What does the needle of a compass point toward?_____

    _____

13. The Earth's (magnetic–geographic) poles are not constant in their position.

14. Matter is made up of tiny microscopic particles called _____.

15. All parts of a magnet show equal strength. (True–False) _____

MP3427 Magnetism and Electricity

Name _____   Date _____

ELECTRICITY AND MAGNETIC FIELDS

In the early 1800s, scientists first discovered that an electrical current passing through a wire was surrounded by a magnetic field. The following activities will illustrate this.

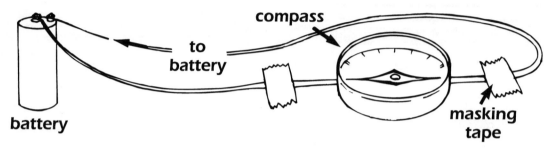

ACTIVITIES

Tape a loop of insulated wire, about 2-feet long, on a flat surface. Place a compass on the wire so the needle will be parallel to the wire as pictured above. Now, connect the battery.

1. Describe what happened to the position of the compass needle. _____

_____

_____

2. Reverse the wires on the battery. Describe what happened to the compass needle. _____

_____

_____

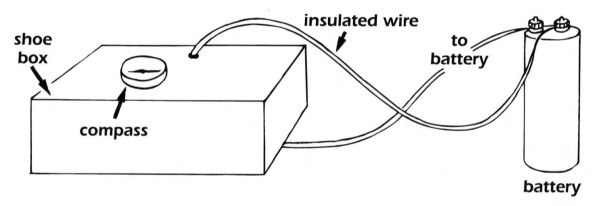

Make a tiny hole in the bottom of a shoe box. Run the wire used in the first activity under the box and up through the hole as illustrated. Connect the wire to the battery. Place a compass near the wire and move it to different positions around the wire. Watch the compass needle. Now, reverse the wires on the battery and repeat the above activity.

3. Describe the behavior of the compass needle. _____

_____

_____

MP3427 Magnetism and Electricity

Name _____   Date _____

MAKING ELECTROMAGNETS

ACTIVITIES

Using a piece of insulated wire about four feet long, make 15 coils around a pencil. Remove the pencil and then tape the coil to a battery.

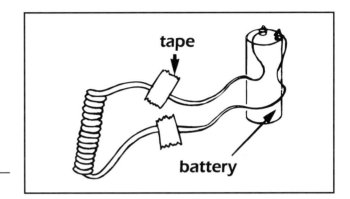

1. Can you detect the north and south poles of the coil by moving a compass around to different positions near the coil?

_____

Insert a 3-inch nail or bolt through the coil to act as a core for the magnetic field. Place a sheet of paper over the coil and core. Connect the battery. You have just made an electromagnet. Lightly sprinkle iron powder over the paper and then tap the paper gently.

2. Can you detect the pattern of a magnetic field? _____

3. Is it similar to the field of a bar magnet? _____

Untape the coil and core, and reconnect the battery.

4. How many thumbtacks will the coil and core pick up? _____

Disconnect the battery.

5. What happens to the thumbtacks when the battery is disconnected? _____

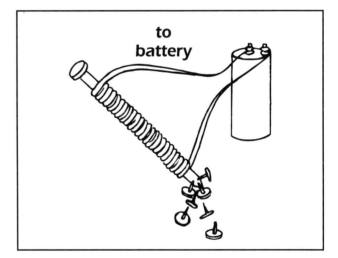

Add 15 more turns of the coil. Connect the wires to the battery.

6. How many thumbtacks will it pick up now? _____
   Why? _____

In the activities above, the nail, or bolt core, became a temporary magnet when an electric current flowed through the coil. The iron core quickly loses its strength once the current is stopped.

Name _____ Date _____

CONSTRUCTING A U-SHAPED ELECTROMAGNET

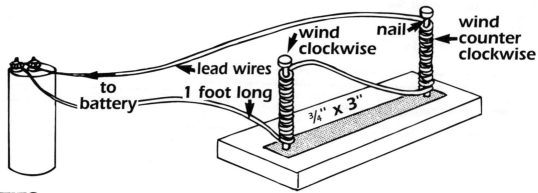

ACTIVITIES

Use a 4-inch piece of wood for the base. Cut a metal strip from a vegetable can 3 inches long and ¾-inch wide. Pound a 2½- or 3-inch nail into each end of the metal strip about ½-inch from the end of the strip. Then, coil 30 turns of wire on each nail as illustrated. Wind one coil clockwise and the other counter clockwise. Connect the lead wires to the battery. Place a compass above one nail and then the other.

1. Which nail head is a north pole? _____ A south pole? _____

2. Turn the magnet on its side. How many thumbtacks will one pole attract? _____
   How many will the other pole attract? _____

3. Compare this number with the number of thumbtacks picked up by the 30-coil iron nail electromagnet you made in the previous activity. _____
   _____
   Explain why. _____
   _____

You can use the U-shaped electromagnet you constructed to make a simple telegraph. Cut a 7-inch by ¾-inch strip from a vegetable can. Tack one end of this strip to one end of a wooden base. Bend the strip so it will be about ¼-inch over each of the nails. Construct and connect the sending key as illustrated. Connect the battery. Press down the key and then release it.

4. What happens? _____
   _____

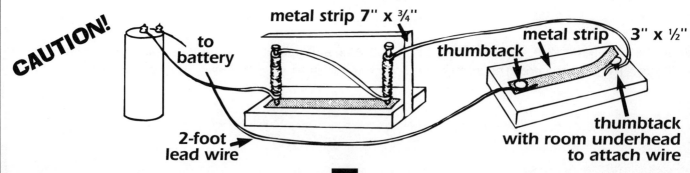

MP3427 Magnetism and Electricity

Name _____   Date _____

MAKING AN ELECTRIC CURRENT DETECTOR

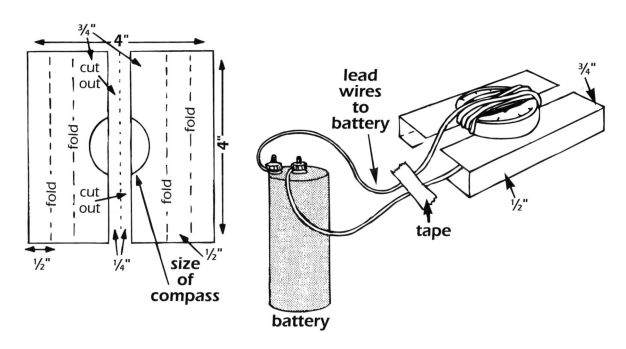

ACTIVITY

Trace the outline of a compass in the center of a 4-inch by 4-inch piece of thin cardboard. Cut out two ¼-inch wide pieces on each side of the center as illustrated above. Make two ½-inch folds on each side of the cardboard. Place the compass over the drawn outline in the center. Using 15 feet of insulated wire, wrap 30 turns tightly around the compass. Leave at least one foot of lead wire at each end. Position the cardboard base so the compass needle is parallel to the wires in the coil. Tape the cardboard base and the lead wires to a flat surface. Your current detector is now ready for use.

Connect the lead wires to the battery.

1. Observe and describe the behavior of the compass needle.

   _____

   _____

   _____

   _____

Now, reverse the lead wires on the battery.

2. Describe what happened to the compass needle when you did this?

   _____

   _____

   _____

   _____

MP3427 Magnetism and Electricity

Name _____ Date _____

MAKING AN ELECTRIC CURRENT WITH A MAGNET

When a coil of wire passes through a magnetic field, a current is caused, or induced, to flow through the wire. The strength of the current will depend on the power of the magnetic field, the number of turns in the coil, and the speed with which the coil passes through the magnetic field.

The giant dynamos that produce electricity for our homes, schools, and industry make use of this magnetic principle.

ACTIVITY

For this activity, you will make a coil to attach to the current detector you made during the previous activity. Cut a piece of insulated wire about 18 feet long. Coil the wire 30–35 turns around a size D flashlight battery. Leave each lead wire at least one foot long. Remove the battery from the coil. Tape the coil in two places to hold it firmly together.

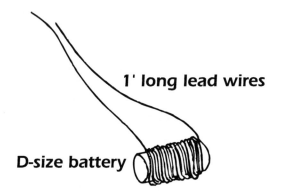

1' long lead wires

D-size battery

Set up the current detector you made. Be sure the compass needle is parallel to the detector coil. Connect the new coil you made to the current detector. The coil should be about two feet from the current detector. Tape the coil-connected lead wires to a flat surface. Now, move a strong bar magnet or one end of a strong horseshoe magnet back and forth through the center of the coil.

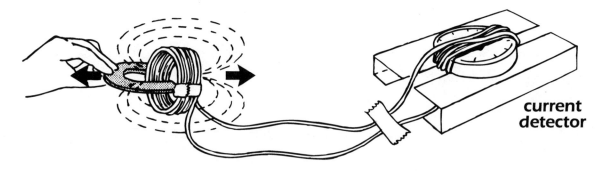

current detector

Observe closely and describe the behavior of the compass needle.

_____

_____

_____

_____

MP3427 Magnetism and Electricity

Name _____   Date _____

STATIC ELECTRICITY AND ELECTRICAL CHARGES

Normally, the atoms within a material are neutral, having an equal number of electrons (–) and protons (+). However, friction between certain materials causes electrons to move from one material to the other. When electrons are added to a material, the material has a negative charge (–). When electrons are taken from a material, the material has a positive charge (+). Electrical charges caused by friction are called static electricity.

Materials with like charges repel each other. Materials with unlike charges attract each other. The following activity will show an example of static electricity.

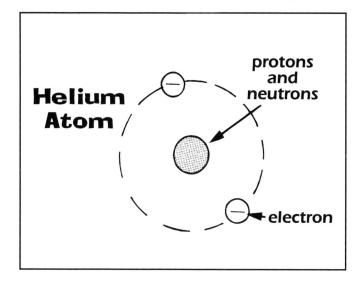

When charges between objects become strong enough, an electrical spark may jump from the (–) charge to the (+) charge.

Lightning is a gigantic spark between two differently charged sections of a cloud or between a cloud and the Earth.

ACTIVITY

Tie strings on two inflated balloons. Rub each balloon with a piece of wool for about 15 seconds. Hold the string ends of the balloons together and allow the balloons to hang freely.

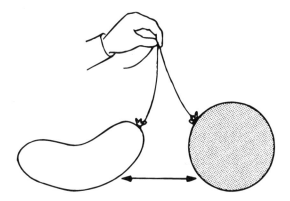

What happens? _____

_____

_____

_____

MP3427 Magnetism and Electricity

Name _____ Date _____

CURRENT ELECTRICITY

Electricity is the flow of electrons through a conductor. The electrons move freely around the nucleus. The **nucleus** is made up of protons and neutrons.

A neutral atom contains the same number of electrons and protons.

ATOMIC STRUCTURE OF A CONDUCTOR (ALUMINUM)

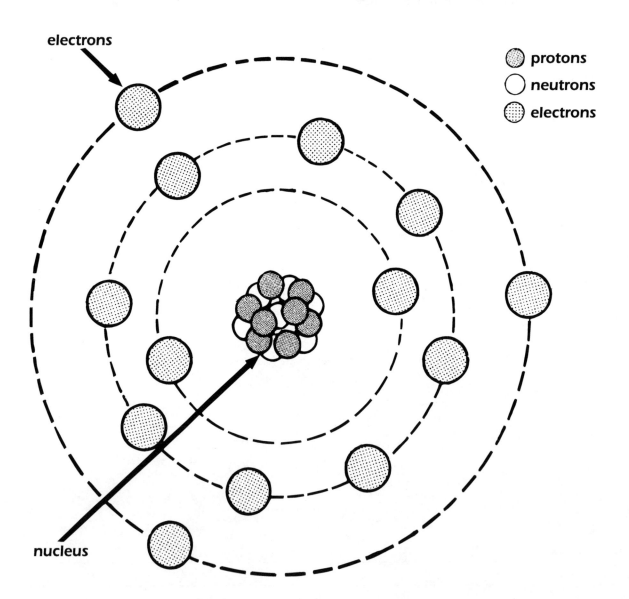

Use the legend to mark the atom. Electrons should have a negative charge (–). Protons are in the nucleus and should show a positive charge (+).

MP3427 Magnetism and Electricity

Name _____   Date _____

WET AND DRY CELLS

A dry cell and a wet cell provide a continuous flow of electrons.

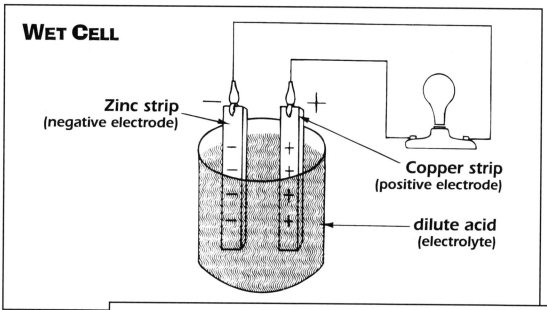

WET CELL

Zinc strip
(negative electrode)

Copper strip
(positive electrode)

dilute acid
(electrolyte)

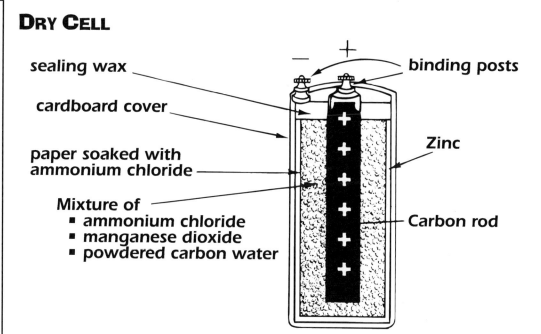

DRY CELL

sealing wax

cardboard cover

paper soaked with
ammonium chloride

Mixture of
- ammonium chloride
- manganese dioxide
- powdered carbon water

binding posts

Zinc

Carbon rod

List three differences between a wet and a dry cell.

1. _____

2. _____

3. _____

MP3427 Magnetism and Electricity

Name _____   Date _____

CONDUCTORS AND INSULATORS

Conductors are materials that allow electricity to easily flow through them. **Insulators** are materials that do not allow electricity to flow through them or offer great resistance to the flow of electricity.

All metals are conductors of electricity. Some are better conductors than others. Silver is the best conductor. Copper is also a very good conductor. It is the metal used in most electrical wiring.

ACTIVITY

Make a simple conductivity tester with a battery, a flashlight bulb, and a socket. Test objects by touching both open ends of wire to the object. The bulb will light up if the material is a conductor of electricity.

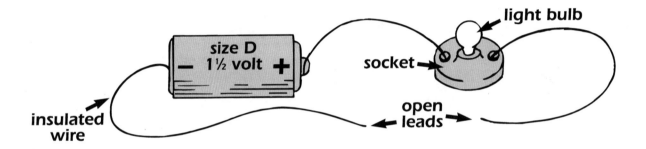

Record your results on the chart below.

Material Tested	Yes	No
wooden pencil		
penny		
aluminum foil		

Material Tested	Yes	No

MP3427 Magnetism and Electricity

Name _____   Date _____

EVERYDAY CONDUCTORS AND INSULATORS

Conductors permit easy passage of electricity. Insulators permit little or no passage of electricity.
Color all of the conductors green. Color all of the insulators red.

POWER CORD

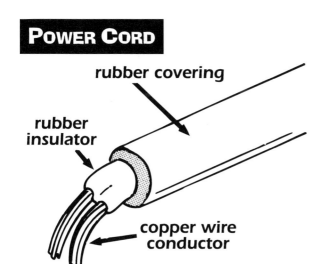

rubber covering

rubber insulator

copper wire conductor

TOASTER

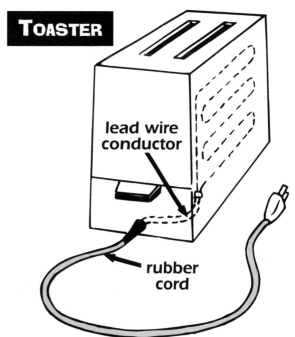

lead wire conductor

rubber cord

UTILITY POLE

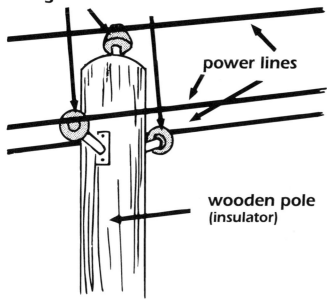

glass insulators

power lines

wooden pole (insulator)

HOUSE WIRING

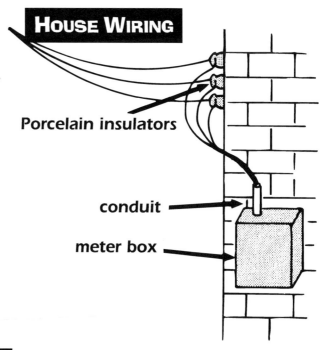

Porcelain insulators

conduit

meter box

MP3427 Magnetism and Electricity

MAGNETISM AND ELECTRICITY

SERIES CIRCUITS

ACTIVITIES

Using a flashlight bulb and a miniature socket, wire the arrangement shown above. Connect a 1½-volt, size D battery.

1. Describe the brightness of the bulb. _____

_____

Add a second socket and bulb as shown. Connect the battery.

2. How has the brightness changed? _____

_____

In a series circuit, the voltage of the battery is divided between the bulbs. Now, unscrew one of the bulbs.

3. Describe what happened to the remaining bulb when you removed one bulb. _____

_____

   In a series circuit, each bulb is dependent on all of the other bulbs for a complete circuit.

Batteries can also be connected in a series arrangement. Using the two-socket connection, connect the battery.

4. Describe the brightness of the bulbs. _____

_____

Add a second 1½-volt, size D battery in a series as shown.

5. How did the brightness of the bulbs change? _____

_____

When batteries are connected in a series, the voltage in the circuit is increased by each added battery. Two 1½-volt batteries act as one 3-volt battery.

24

MP3427 Magnetism and Electricity

Name _____ Date _____

PARALLEL CIRCUITS

ACTIVITIES

Connect a miniature socket and flashlight bulb as shown. Connect a 1½-volt, size D battery.

1. Describe the brightness of the bulb. _____

_____

Now, add a second socket and bulb.

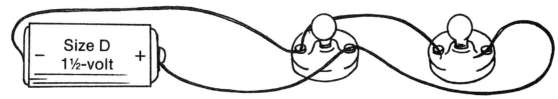

2. How did adding the second bulb to the circuit change the brightness of the first bulb?_____

_____

   In parallel circuits, each bulb receives the full voltage produced by the battery.

While you have both bulbs burning, remove one bulb from the socket.

3. What happens to the other bulb?_____

_____

   In a parallel circuit, each bulb operates separately from the others.

Batteries can also be connected in a parallel arrangement. Use the 2-bulb parallel wiring. Connect the 1½-volt, size D battery. Again, observe the brightness of the bulbs. Now, add a second 1½-volt, size D battery in parallel connection as shown.

4. How did the brightness change? _____

_____

Two 1½-volt batteries connected in parallel will produce only 1½-volts for the circuit. They will, however, cause the bulbs to burn twice as long before the batteries wear out.

MP3427 Magnetism and Electricity

Name _____   Date _____

KINDS OF CURRENT

Current is the movement of electrons from one atom to another.

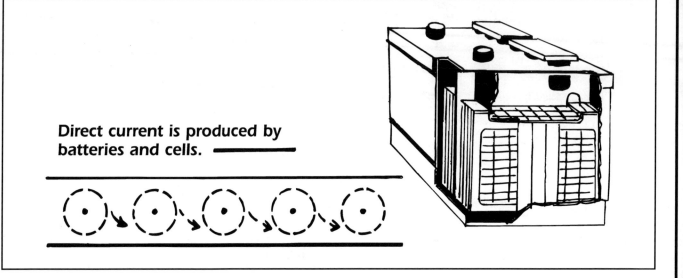

Direct current is produced by batteries and cells. ——————

Alternating current is produced by dynamos.

List five things that can be found at home that use alternating current.

1. _____
2. _____
3. _____
4. _____
5. _____

List five things that you can buy that are run by direct current.

1. _____
2. _____
3. _____
4. _____
5. _____

26

Name _____   Date _____

PARALLEL VERSUS SERIES

The wiring in our house makes use of parallel circuits.

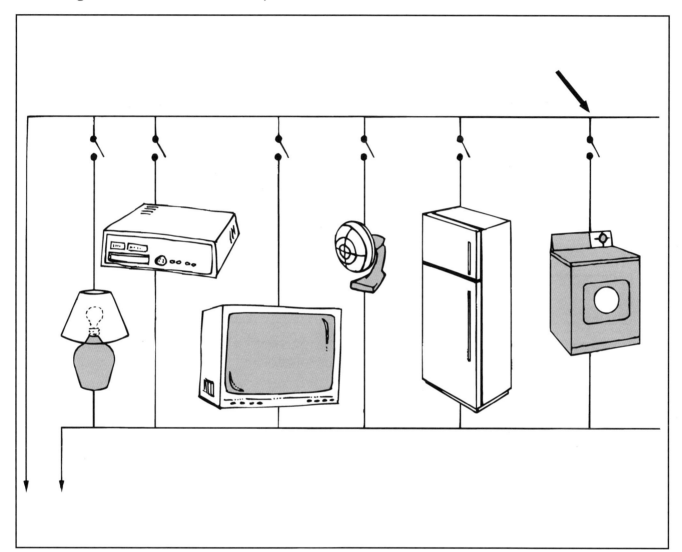

If your house were wired in series circuits instead of parallel circuits, what would happen when a light bulb burned out?

_____

_____

_____

_____

Name _____   Date _____

REVIEW OF ELECTRICITY

Complete the following.

1. Static electricity is an electrical charge caused by_____.

2. In a series circuit, each bulb (is dependent on–acts separately from) the other bulbs in the circuit.
   _____

3. The strength of an electrical current caused when a coil of wires cuts through a magnetic field depends on:

   a. _____

   b. _____

   c. _____

4. Our homes are wired in (parallel–series) circuits.

5. An electrical current flowing through a wire is surrounded by a _____.

6. Wood is a good conductor of electricity. (True–False) _____

7. Neutral atoms have an equal number of _____ and
   _____.

8. Static electricity is of great use to man. (True–False) _____

9. When you connect several batteries in a series, the voltage (stays the same–increases).

10. An electromagnet core is a temporary magnet. (True–False)

11. An electrical current is the flow of _____ from one place to another through a wire.

12. A toaster is an example of _____ energy being changed into heat energy.

13. Lightning is an example of _____ _____.

14. For man to make use of electricity _____ is needed.

15. All metals are conductors of electricity. (True–False) _____

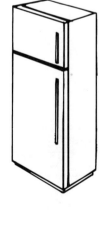

MP3427 Magnetism and Electricity

MAGNETISM & ELECTRICITY BACKGROUND MATERIAL

Pages 1–2: Follow the page 1 activity with a discussion of student answers. Discuss page 2. All items pictured make use of permanent artificial magnets. Students may suggest other items that may or may not make use of permanent magnets. Students should research their suggestions for accuracy.

Pages 3–4: Special materials needed: one bar magnet, one horseshoe magnet, iron filings or iron powder. Small pieces cut from a ball of steel wool are a possible alternate to iron powder or iron filings. Students will see that magnetic field patterns arc from north to south.

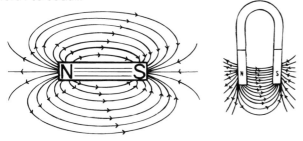

Page 5: Special materials needed: two bar magnets and two horseshoe magnets.

Page 6: Special materials needed: one bar magnet, one horseshoe magnet, two 3" common iron nails, iron powder or iron filings.

Page 7: Special materials needed: 12 to 15 assorted items such as found in junk drawers in the kitchen or basement. Follow this activity with a discussion of results. There may be disagreement in results because some metal items are made of materials that are not magnetic. Nails, screws, washers, and paper clips are examples of items that are usually made of magnetic materials, but are occasionally made of non-magnetic metals such as aluminum.

Page 8: Special materials needed: one bar magnet, pieces of paper, plastic wrap, cloth, aluminum foil, pieces of rubber sheeting about 4"; a ceramic saucer or plastic dish; several thumbtacks; thin piece of styrofoam about 1" by 1"; pieces of glass pane; cardboard; pieces of styrofoam 3" by 3" or larger; flat piece of metal from a can about 4" by 4". (Egg cartons and meat trays are a good source of thin styrofoam.) After removing both ends from a can, the

remaining cylinder can be easily cut with a tin snip. Rubber sheeting may be cut from a balloon. CAUTION! On student pages, remind them to work carefully and safely with sharp materials.

Pages 9–10: Special materials needed: one medium ceramic or plastic bowl; two needles; 1" x 1" piece of thin styrofoam; one bar magnet; thread; masking tape; ice cream stick; glass jar. The method for magnetizing the needle is explained in the activity on page 6. **Note:** you may have better results with these activities if the work is done on a wooden table.

Use page 10 as a follow-up for discussion. Have students research the location of the magnetic poles and compare their current location with their position in past geologic periods.

Pages 11–12: Special materials needed: two 3" common iron nails; pair of pliers; propane burner; matches. A candle flame is a possible alternate to the propane burner, but the effect is not as dramatic. If a candle flame is to be used, smaller nails should also be used. The nail should be held in the candle flame for 3 to 4 minutes. The nail should be moved slowly so all parts of the nail are equally heated. USE EXTREME CAUTION!

Follow this activity with page 12. In addition to the points illustrated on the page in support of the molecular theory of magnetism, one additional point is to be made. Heating and pounding disarranges the molecules in an iron magnet causing the magnet's force field to become weaker.

Page 13: See answer key. Review the answers with students to reinforce correct concepts.

Page 14: Special materials needed: a 1½-volt battery; compass; piece of insulated wire 2–3' long; masking tape; a shoe box or tissue box. (Insulation should always be stripped from ends of wires that are connected to batteries or to each other.)

In the activities on this page, the compass needle will move from a parallel position to a position forming 90° angles to the wire. Students may describe this in their own words or may draw a picture. **Note:** A 6-volt battery may be substituted for the 1½-volt battery in any of the electromagnetic activities.

Page 15: Special materials needed: piece of insulated wire 4' long; pencil; one 3" common iron nail; a 1½-volt battery; iron powder or filings; iron thumbtacks; masking tape. Pattern of magnetic field will appear similar to, though not as clear, as picture of bar magnet shown for pages 3 & 4.

Page 16: Special materials needed: piece of insulated wire 4' long; two 2½" or 3" common iron nails; three metal strips from a vegetable can—one 3" by ½", one 3" by ¾", and one 7" by ¾"; one 1½-volt battery; tacks; two pieces of wood 2" wide and 4" long; compass.

The two iron nail coil cores are connected by the metal strip which, in effect, gives each nail core the added strength of the other coil.

Page 17: Special materials needed: piece of thin cardboard 4" by 4" and about ¹⁄₁₆" thick; one 1½-volt battery; compass; piece of insulated wire 15' long; masking tape. Use with page 18.

Page 18: Special materials needed: one strong bar magnet; one strong horseshoe magnet; 18' of insulated wire; masking tape.

Pages 19–21: Special materials needed: two rubber balloons; a piece of wool or fur; string. Note: All static electricity activities work best on dry, cool days. High humidity may cause poor results. The chemicals in both dry and wet cell batteries cause the zinc material to dissolve. As each zinc atom enters into solution, it leaves two electrons remaining on the zinc strip or lining. This creates an excess of electrons on the zinc plate giving it a negative (–) charge.

Pages 22–23: Special materials needed: one 1½-volt battery; one flashlight bulb (screw type); one miniature socket; three 1' long pieces of insulated wire; varied materials for testing. Materials collected for activity on page 7 may also be used for this activity. Point out that without the use of insulators (nonconductors), it would be impossible to guide high voltage electrical current from one place to another. In our homes, insulators help protect our families from the dangers of electrical fires and electrical shock.

Page 24: Special materials needed: two 1½-volt batteries; two miniature sockets and flashlight bulbs; three pieces of insulated wire 1' long. As an extra activity at the end of this page, have students reverse the batteries to see what happens. (Bulbs will not light.) **Note:** A student-made battery holder for connecting 2 batteries may be constructed with the use of rubber bands, a toilet tissue cylinder, and—at the ends—a twisted piece of copper wire with 6" lead. A single battery can be wired for use in a similar manner.

By now, students will have discovered that the wiring gets hot when the current flows through it. By forcing electricity to flow through materials offering some resistance to the flow of electricity, we can convert electrical energy to heat energy.

Pages 25–27: Special materials needed: two 1½-volt batteries; two miniature sockets; three pieces of insulated wire, 1' long. Homes and other buildings have a number of parallel circuits in them, each protected by a fuse or a circuit breaker. **Note:** Large 1½-volt dry cells with screw-type terminals can be obtained from many hobby shops. They will last much longer, but are more expensive.

Page 28: Be sure to review the correct answers with students as soon after the test as possible.

MP3427 Magnetism and Electricity

ANSWERS

Page 1
Answers will vary.

Page 2
hooks for hanging utensils
electric can opener
speakers in radios, stereos, televisions, etc.
lawn mower with pull-type starter
magnetized seal on refrigerator/freezer doors
small toys using battery motors
cabinet door fasteners

Page 3

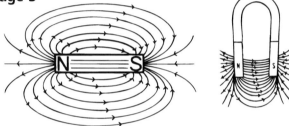

Page 4
Student drawings will show that more iron particles collect around poles of the magnet.

Page 5
Student responses will vary but should include the following: Like poles push each other away (repel). Unlike poles pull toward each other (attract) and cling together.

Page 6
1. The nail attracts more iron powder than before.
2. The number of strokes increases the strength of magnetism of the nail.
3. Yes
4. No—or weaker

Page 7
Answers will vary depending on items selected to be tested with the magnet. Generally, most metallic items will be attracted to the magnet; nonmetallic items will not.

Page 8
1. Aluminum foil
2. The thumbtack follows the magnet.
3. No
4. The thumbtack follows the magnet.

5. Answers will vary.
6. No

Page 9
The needles are in a general north-south direction.

Page 10
The matching poles would repel each other.

Page 11
1. Answers will vary.
2. Answers will vary.
3. The nail that is not heated.

Page 12
1. Yes
2. The field is reversed.
3. Yes
4. The picture would show the reverse of the molecules in the box at the top on the right.

Page 13
1. True
2. False
3. The poles are the areas of a magnet having the greatest strength.
   They are located at the ends.
4. True
5. The area in which a magnet exerts its strength (force).
6. unlike, Like
7. False
8. a) Stroking the nail with another magnet.
   b) Placing a nail on or near a magnet so the lines of force will travel through the nail for a period of time.
9. True
10. True
11. True
12. The magnetic north and south poles
13. magnetic
14. atoms
15. False

Page 14
1. The poles align with the battery.
2. The poles reverse position.
3. The poles reverse position but they remain at 90° angles to the wire.

Page 15
1. Yes
2. Yes
3. Yes
4. Answers will vary.
5. Thumbtacks will drop from the nail.
   Answers will vary but the number will be greater because the power is increased.

Page 16
1. Answers will vary.
2. Answers will vary.
3. U-shaped magnets pick up more/stronger magnets. Metal strip is pulled down to nail heads and then released.

Page 17
1. The needle will swing from a parallel to the coil to 90° position to the coil.
2. Poles reverse position but remain at 90° angles to the coil.

Page 18
Needle swings to both sides of parallel position.

Page 19
The balloons repel each other.

Page 22
Answers will vary.

Page 24
1. Answers will vary.
2. Bulbs burn at a dimmer brightness.
3. The remaining bulb goes out.
4. The brightness is increased in both bulbs.
5. Brightness stays the same.

Page 25
1. Answers will vary.
2. Brightness stays the same.
3. It remains lit.
4. Brightness stays the same.

Page 27
The electrical equipment would go out .

Page 28
1. friction
2. Is dependent on
3. a) the strength of the magnetic field
   b) the number of turns in the coil
   c) the speed the coil cut through the magnetic field
4. parallel
5. magnetic field
6. False
7. electrons, protons
8. False
9. increases
10. True
11. electrons
12. electrical
13. static electricity
14. steady flow of electrical charge, or current
15. True